CREATION,
EVOLUTION and
EXTINCTION

CREATION, EVOLUTION and EXTINCTION

Matter is neither created nor destroyed; it is merely changed in form.

ESSAY

Robert A. Moore, BS

iUniverse, Inc.

New York Lincoln Shanghai

CREATION, EVOLUTION and EXTINCTION
Matter is neither created nor destroyed; it is merely changed in form.

iUniverse, Inc.

For information address:
iUniverse, Inc.
2021 Pine Lake Road, Suite 100
Lincoln, NE 68512
www.iuniverse.com

ISBN: 0-595-32727-3

Printed in the United States of America

Contents

Synopsis of This Essay

This essay was written during the year of 2003, in celebration of the fiftieth anniversary of the discovery of the double helix structure of DNA. This discovery allowed for the mapping of the twenty-three pairs of chromosomes that make up the human genome for our species, Homo Sapien. This essay is also written for every person who grew up in a religious environment; and, when the time came, moved on to the study of higher education with its new knowledge of science, biology, life, and evolution. This new knowledge brought on doubt and guilt about our religious upbringing.

People felt they had to make a choice between believing in creation or believing in evolution, when actually, creation and evolution are one and the same. Our God created each and every species along with evolution, which is the process for maintaining that species in the many different types of environments of the world. That power includes the extinction of many of those species when they can no longer compete. Everything on earth is under God's control. God is all-powerful! God gave our species the ability to think and to make choices, no other species has this gift and he wants us to gain wisdom as outlined in the Book of Proverbs.

Each species in the universe has its own genome, its individual set of DNA instructions and with its own special number of chromosomes in order to reproduce its own kind. For this reason, no species can successfully breed with a different species and reproduce its kind. It may, however, produce a hybrid. As stated, the human genome has twenty-three pairs of chromosomes. Large apes, including gorillas, chimpanzees, baboons, etc. has a genome with twenty-four pairs of chromosomes, therefore, it would be impossible for our two species to reproduce and form a new species. In the case of a donkey, with 31 pairs of chromosomes; and a horse, with 32 pairs, they can mate and produce a mule, which is a well-developed and strong animal, but the mule cannot reproduce with another mule. (The mule gets 32 chromosomes from its father and 31 from its mother.) It is thought that a male Homo Sapien could have had sex with a female Neanderthal, (at this time the number of chromosomes is not known for the Neanderthal), and the hybrid offspring would be strong and healthy, but it is believed it could not reproduce. Therefore, it could not create a new species.

The discovery and use of DNA has proven that Charles Darwin was correct as far as altering each species, depending on its environment, with his theory of evolution for each species. Evolution, mutation of the genes, maintained each species and allowed that species to survive in the different environments of the world. However, Charles Darwin was never specific about the beginning of each new species, and there had to be a first, He said it just happened. This is where God and creation of new species arrives in the plan. Each new species must be created since each new species has its own genome, its own number of chromosomes. Ever since "The Origin of the Species" was first published, the evolutionist and the creationist have been fighting each other as if this was an either/or situation. Do not sell God short, he is involved in all things scientific as well as moral. After all, this is my Father's world.

No one can prove, nor disprove, the existence of a superior being, a God. That depends on your own individual belief and faith. Stephen Hawkins, in his book, "A Brief History of Time", while discussing the Big Bang theory and the Expanding Universe theory of Edwin Hubble, says, "An expanding universe does not preclude a creator, but it does place limits on when he might have carried out his job!" Most of the great scientist of the past, Copernicus, Galileo, Newton, and Einstein, believed in God.

The earth was created much later than the universe. It is possible, but not probable, that God could have created the planet earth to use as a giant laboratory for his experiments to create different species and to use evolution to maintain, in our example, the various races of the same species. If an experiment got out of hand, example with dinosaurs, he would cause a mass extinction. All species eventually become extinct and something new replaces these older species.

I believe that God gave our species a soul about fifty thousand years ago, at a time known as, "The Mind Change", or "The Mind Bang", and there had to be a first of our species that had a soul and that first was called Adam and Eve. No other species has the ability to reason and to make choices and there is, at the present time, only one human species and that one is our species, Homo Sapien. This ability was a gift to man from God. God gave man a soul. The body evolves, the soul can live on, if you believe and obey God's laws.

God gave us rules to live by where we learn to live together. These rules are "The Ten Commandments" and if you live by these rules your soul will live forever. I believe that it does not matter how long ago God created the earth, or how he created the earth, or if the Bible is to be believed word for word. The only thing that matters is how you live with your fellow beings by obeying God's commandments and the teachings of Jesus Christ. You may not agree with my premises, nor my belief, that is for you to decide and God encourages you to think for yourself. Please read the Book of Proverbs.

Belief in God brings a chance for the Soul to live forever, if you live by the rules. To do otherwise is to have no hope for a future and eternity is a long, long time. To do nothing, to live in the past, to not use the mental gifts that God gave you, is not following God's teachings. Believe what you will and let others believe what they will, but follow the Commandments.

"High up in the north in a land called Svithjod, there stands a rock. It is a hundred miles high and a hundred miles wide. Once every thousand years a little bird comes to the rock to sharpen its beak. When the rock has thus been worn away, then a single day of eternity will have gone by." Hendrik Van Loon.

Preface

This essay is a summary of many ideas developed in the past fifty plus years about the Universe. The theory of "The Big Bang" and the "The Big Crunch". The beginning of the Earth, its geological time periods, and life. The Periodic Table with the elements. Mass extinction of some species and the carbon dating of past living material. Our own species genome, RNA, DNA, mitochondrial DNA, pre-historic humans, our species Homo Sapien, God, and the "Mind Change", the chronology of man, where we came from and where we are going.

This essay is not being written as a scientific paper. The majority of ideas outlined in this work are the thoughts and words of many different people; many quotations are linked together by a layman with a scientific background who is eager to give credit to all of the proper people. There is a bibliography of all of these authors at the end of this essay in hopes that the reader will purchase the books that interest them.

There are things that we will never know with certainty. So don't jump to conclusions that may lead to hopelessness for the future. As Albert Einstein said, "God does not play dice." Neither should we gamble with our future because our minds are closed. Do as God ask us to do in the Book of Proverbs, continue to learn. Let's take a look at where we are and how we got here. Keep your mind open at all times; let it be open 24 hours a day.

We can make some assumptions and then decide whether we agree or not, that is our choice as a living, thinking human being.

1. God created the universe with the "Big Bang" ten to twenty billion years ago.

2. God created our solar system four to five billion years ago.

3. God created every living thing on earth and maintains them in a system called evolution. The proof of these lies in the study of RNA and DNA and according to laws of heredity as set down by Mendel.

4. God gave a soul to modern man, at some particular time in our past, and it doesn't really matter exactly when. This was to give man the ability to think and to make choices on his own. No other species has this gift or can make these choices.

5. The body is evolution. The soul is God given and it's future depends on what you believe and how you act as a moral human.

6. You can very easily live your life on earth with a belief in God, the teachings of Jesus Christ, creation and evolution, and by using the Ten Commandments as your guide. Do so and your soul will live forever. It is important, however, that you let other people think for themselves. God gave them a choice too.

7. If some people do not believe in God, then it is also important that you allow them to think whatever they want, that is their choice. We can only hope they follow at least four of the Ten Commandments even if they do not believe in God. Those are, Thou shall not kill. Thou shall not steal. Thou shall not bear false witness. Thou shall not covet thy neighbor's property. I believe that if you do not believe in God you will have no future.

We are going to look at all of these subjects and then make up our minds based on what we believe.—"You are what you believe."

The goal for the rest of this essay *is to keep it simple*.

1

The Big Bang[1]—The Big Crunch[2]

This chapter is written as I would write a book report in school. The report is on a book written by Stephen Hawking, "A Brief History of Time" using the updated and expanded tenth anniversary edition published by Bantam Books. I highly recommend this book for your personal library.

Stephen Hawking is a wonderful man who has suffered over the past, almost forty years, from the debilitating effects of Amyotrophic Lateral Sclerosis, also known as "Lou Gehrig's" disease. This disease has confined Stephen to a wheel chair for many years and he communicates by using a computerized voice synthesizer. "Stephen Hawking was born in 1942 on the anniversary of Galileo's death. He holds Isaac Newton's chair as Lucasian Professor of Mathematics at the University of Cambridge in England. Widely regarded as the most brilliant theoretical physicist since Einstein." His latest book is "The Universe in a Nutshell", which I also recommend for your personal library. Stephen Hawking is truly a remarkable man.

From Chapter 1 of "A Brief History of Time", Page 9 of my copy.

---

1.   It should be mentioned that the term, "Big Bang" was coined by Fred Hoyle in 1950 during a radio series, "The Nature of the Universe", by the British Broadcasting Corporation (BBC).
2.   The Big Crunch is a term used for the end of the universe in case the galaxies ever reverse the Big Bang and have it go back the other way causing the Big Crunch. This could possibly result in the beginning of a new Big Bang and the start of a new time period. This could go on forever.

"In 1929 Edwin Hubble made the landmark observation that wherever you look, distant galaxies are moving rapidly away from us. In other words, the universe is expanding. This means that at earlier times objects would have been closer together. In fact it seems that there was a time, about ten or twenty thousand million years ago, when they were all exactly the same place and when, therefore, the density of the universe was infinite. This discovery finally brought the question of the beginning of the universe into the realm of science."

"Hubble's observations suggested that there was a time, called the big bang,"

"One can imagine that God created the universe at literally any time in the past. On the other hand, if the universe is expanding, there may be physical reasons why there had to be a beginning. One could still imagine that God created the universe at the instant of the big bang, or even afterwards in just such a way as to make it look as though there had been a big bang. An expanding universe does not preclude a creator, but it does place limits on when he might have carried out his job."

2

The Earth—Life Begins

There are two Geological Charts shown on the following two pages, as Figure 2-1 and 2-2, please refer to these charts while reading this section.

The first several billion years there was no life on earth. The earth was cooling off, rocks and crust were forming. This "Big Bang" part, of the geological chart, used to be called the Azoic Era, but now it is rarely used. In Figure 2-1, from the year 1961, the information used listed both the Archeozoic Era and the Azoic Era. Between then and now all of the very early time has been referred to as the Precambrian Period. From this primordial stew single cell anaerobic bacteria began. Single cells lasted for millions of years. At this time, Figure 2-1, the earth was thought to be 4.5 billion years old, in the meantime, a short fifty some years, science has pushed this period back another three million years to 4.8 billion years, see Figure 2-2. Many changes have been made in the dating of materials and using other guidelines to determine the different time periods, among them is the dates of various extinction, the last of which was at the end of the Permian Period. This extinction did away with the dinosaurs among others. I have never seen two different geographical charts with the same information, and the differences are in millions of years. Do not let this trouble you as most of you could care less. The important part of this essay is not billions of years ago but just the past of our own species, or about 50,000 years, give or take a few thousand years.

The Mesozoic Era with its three periods, Triassic, Jurassic, and Cretaceous was the era of the dinosaurs until they were wiped out with the fifth mass extinction. They ruled the earth for 140 million years. The removal of the dinosaurs allowed for the evolution of modern animals, monkeys, and apes. This era began 65 million years ago. Some of the geological charts have time periods different from chart to chart. But, what is a few million years between friendly scientist. If you happen to be a fundamentalist and you ask me, "Where you there?" I will have to admit that I was not there, but because I have an open mind I did not have to be there, They left their tracks with fossils, and more recently, we can read their DNA. (Deoxyribonucleic Acid)

At first all life was living in water, there was no life on land. Land was barren except for rocks and minerals lasting for eons. Then God caused plants to grow on the land. These plants could be used for food, so some of the fish adapted to land over millions of years so they too could eat the food in a safer environment. That's the way evolution works. The fish evolved into amphibians, thus to reptiles, dinosaurs, etc.

While the dinosaurs were living on earth the larger mammals and man could not survive and the dinosaurs lived for about 140 million years. Perhaps God got tired of them, anyway a large meteor wiped them all off the earth. Some of the small animal survived and were having a ball until some new larger predators could evolve. New species of smaller birds were evolving along with the first prehistoric horses. Then about 35 million years ago the monkey, the small ape, and some more modern looking animals came along. God must have liked the monkey, as they were fun to watch. Man was to be one of the last species, and one of the most successful, for God to create.

The idea behind this essay of geological time, and covering billions of years, hopefully will whet your appetite for additional knowledge. The Geological Chart shown in figure 2-1 is about sixty years old while figure 2-2 is less than ten years old.

It is easy to understand how the people who wrote the Bible would be a long way off with regard to time, and interpreting time. As late as the eighteenth century, it was commonly believed that the universe itself had been created less than 6,000 years ago and life a few days later. In the Middle Ages, Archbishop James Ussher proclaimed that the earth was created at 9 AM, October 23, 4004 BC as a result of his study of the Book of Genesis. Some people still believe this to be true; and it is true for the time for the beginning of Hebrew history. In 4004 BC, the early stories of the Bible would have had to be handed down, father to son, by word of mouth because writing would not be invented for another thousand years. Writing was invented in Egypt and Mesopotamia around 3,000 BC, or 5,000 years ago. (Just another thought, the concept of a "zero" and a decimal system was not common knowledge until 1100 AD, which would mean that numbers were not easy.)

For anyone who believes in God and Jesus Christ, what possible difference could a mix up in time or numbers, effect your belief in, "The Ten Commandments" or, "Do unto others as you would have them do unto you."

GEOLOGICAL PERIODS* FIGURE 2-1

ERA	PERIOD	BEGIN (Mil of yrs. Ago)	DUR AT-ION	EPOCH	LIFE BEGAN
CENO-ZOIC	Quaternary		Till now	Recent	Stone Age begins
		1	1	Pleistocene	Ice Age—Prehistoric Men
		12	11	Pliocene	1st Ape like humans.
		28	16	Miocene	Mammals—Modern.
		35	7	Oligocene	First monkeys & apes.
		50	15	Eocene	Primitive Horse.
	Tertiary	60	10	Paleocene	Dinosaurs are extinct.
MESO-ZOIC					
	Cretaceous	130	70	Flowering plants	Strong mammals
	Jurassic	170	40	Dinosaurs peak	Birds
	Triassic	200	30	Mammals	Dinosaurs flourish
PALEO-ZOIC					
	Permian	230	30	Dinosaurs begin.	Will last 140 mill. Yrs..
	Carboniferous	290	60	Reptiles	Seed plants
	Devonian	330	40	Sharks-Insects	Amphibians
	Silurian	360	30	First jawed fish	Plants
	Ordovician	420	60		Jaw-less Fish

	Cambrian	500	80	Multi-cellular	Mollusks—Anthropoids
PROT-ERO-ZOIC					
	Keweenawan	750	250	Lava flows in the	Great Lakes region. Life abundant in sea.
	Huronian	1000	250	Sediments/iron	In Lake Superior.
	Timiskaming	1300	300	Limy sea plants.	A few fossils-no other living things known.
ARCHEO ZOIC					
	Keewatin	2100	800	Limy sea plants.	Probably algae.
AZOIC					
	BIG BANG	4500	2400	No life on Earth.	Cooling, forming crust.

• **Information from: "The World Book Encyclopedia"**
Copyright 1961

GEOLOGICAL PERIODS* FIGURE 2-2

ERA	PERIOD	BEGIN (Mil of yrs. Ago)	DURA T-ION	EPOCH	LIFE BEGAN
CENO-ZOIC	Quaternary	1	Till now	Holocene	Stone Age begins
		1.8	0.8	Pleistocene	Ice Age—Prehistoric Man-mammoths
		5	3.2	Pliocene	1st Hominids
		23	18	Miocene	1st Ape like humans
		34	11	Oligocene	First Apes.
		57	23	Eocene	First Horse, 1st monkey
	Tertiary	65	8	Paleocene	Dinosaurs are extinct.
MESO-ZOIC					
	Cretaceous	144	79	Flowering plants	More mammals
	Jurassic	208	64	Dinosaurs peak	First Birds
	Triassic	245	37	First Mammals	First Dinosaurs
PALEO-ZOIC					
	Permian	286	41		
	Carboniferous	360	74	Reptiles	Seed plants
	Devonian	408	48	Sharks-Insects	Amphibians
	Silurian	438	30	First jawed fish	Plants
	Ordovician	505	67		Jaw-less Fish

Cambrian	550	45	Multi-cellular	Mollusks—Anthro-poids
Proterozoic	2500	1950	Bacteria	Green Algae
Archean	3800	4000	First Life	Bacteria—Oldest Rocks
BIG BANG	4800	1000	No life on Earth.	Cooling, forming crust.

• **Information from: "The Book of Life", Introduction**
Life and Times 1993 by Michael Benton

3

Prehistoric Hominids

Prehistoric Hominids are by definition any of the family (Hominidae) consisting of two legged primates including all forms of early humans.

Hominids go back in time as far as two, to possibly three, million years ago. It is debatable, depending on which paleontologist you read. However, most of them are known as Homo Habilis, Australopithecus, Heidelberg Man, Homo Erectus, Java Man, Peking Man, Neanderthal Man, Cro-Magnon, and early Homo Sapiens. Perhaps you have heard of Lucy the name given to fossil bones found in Ethiopia in 1974 by Donald Johnson and Tom Gray. Lucy is said to have lived about 2.3 million years ago and was an adult woman about 25 years old. Evidence from her bones indicated she was bipedal, although there was reason to believe she was also partly arboreal (tree-dwelling). Her species was assigned to Australopithecus afarensis. None of these species, including early Homo Sapiens, had the ability to think, to reason, or to make choices. They all reacted to instinct. You may also have heard of the term, LUCA, this is a set of letters meaning "Last Universal Common Ancestor" and represents a connection of all the early Hominids yet to be discovered.

It is not known at this time how many chromosomes these prehistoric species had, but there was an attempt to group these Hominids into special bone structures over the years. Especially, by measurements of the human skull, by brain capacity, etc. Many scientists have tried to classify humans by their skeletal construction. Also, the strict evolutionist would like other groups of primates, consisting of monkeys, apes, of gorillas, chimpanzees, and baboons to be included. These primates have 24 pairs of chromosomes, one set from the father and the other set from the mother for a total of 48 chromosomes, making this group, of monkeys and apes, a different species from our species. Homo Sapien, an entirely different species, has 23 pairs of chromosomes, for a total of 46 chromosomes.

So, for this reason, we know that they could not have bred successfully with the monkeys, apes, etc. We do not know how many sets of chromosomes other hominids had but, it is believed that our species could not have successfully bred with, for example, Neanderthal. (To this date there is no DNA confirmation).

All of the prehistoric Hominids humans have become extinct except for Homo Sapiens. The last group to become extinct was the Neanderthal. Their span on earth lasted from about 200,000 years ago to about 28,000 to 30,000 years ago. Homo Sapiens began in Africa about 150,000 years ago and for their first 100,000 years they reacted to instinct alone, they were like all other human hominids. It is thought by some scientist that a "Mind Bang" occurred approximately 50,000 years when God gave man a soul with the power to think and to make choices. Since about 45,000 years ago, or, when our species reached Europe, they coexisted with the Neanderthal for a period of about 28,000 years. There is no record of mistreatment of the Neanderthal during this time. However, it is possible the Neanderthal learned like a dog from Homo Sapiens, like using tools or burials, etc. but he could not compete.

4

Extinction

In the beginning God created the Universe with the "Big Bang" between ten to twenty billion years ago. Some four or five billion years ago God also created our solar system using stardust and hydrogen. After millions of years for cooling off, God could have selected the Earth, which was at that time void, to create the first single cell of life. This was the beginning of a long, long experiment in the evolution of these cells, either singular or later multiple, to learn what glorious forms of species could be produced. God often changed his mind and would cause some cells or species to become extinct. This could have been God's laboratory for the rest of the universe.

Extinction is a normal part of evolution. Many times in the history of life there have been mass extinction to clear off the playing field and start over. The fourth mass extinction, around 250 million years ago, allowed the beginnings of the reptiles and dinosaurs to take hold. They grew to enormous proportions and the only animals left alive were small nocturnal animals that could hide in the day. Even birds, which developed in this period, were very large and mostly scavengers used to clean up the very large reptiles. After 140 million years in control of the world, God decided to clear the playing field again with the fifth mass extinction, around 65 million years ago. So he had a comet bang into the earth and kill the entire dinosaur species as well as the large birds and other species related to the dinosaur's existence.

Extinction is also an important part of evolution. Species begin, live, make subtle changes, then, for a myriad of reasons, they die out and become extinct. The average time for the existence of an animal species is thought to be about four million years; for some it is less, for others much longer but enough to be a regular occurrence. It has been estimated that some ninety to ninety-five percent of all the animal species that have ever lived on earth are now extinct either from normal evolution, or from a catastrophic mass extinction.

There are many causes for extinction: the climate can make drastic changes over long periods of time; new predators, or ones with different adaptations, may come along and reek havoc; or, a geological movement of plates causing the shift of continents, or mountains, can be catastrophic. All of these things can cause species to become extinct. David Raup and J. John Sepkoski of the University of Chicago have suggested in 1983 that perhaps mass extinction has followed a periodic cycle, with spacing of 26 million years between events over the past 250 million years. There are others with different suggestions and theories. Michael Benton says, "Twenty or more global mass extinction have been identified, some more convincing than others, during the history of life." (From "The Book of Life".)

In addition to regular individual species extinction, there has been, at least, five major mass extinction's over periods of millions of years since life began. A mass extinction, like the fifth one, wiped out the dinosaurs, can level the playing field making it possible for entire new species to advance including the ancestors of man. It is thought that a large asteroid or comet hit the earth some sixty million years ago, at the end of the Cretaceous Period. There was also a mass extinction, the fourth, at the end of the Permian Period, which cleared the way for the beginning of reptiles, thus the dinosaur species. The dinosaurs ruled the earth for over 140 million years. While they were here only small animal species could survive by hiding in holes in the earth and only coming out at night. Man and modern animals would not have a chance for survival by competing with the dinosaurs.

The way modern man and animals have come about, there are those who say we are now the cause of the next mass extinction due to global warming, a continuing epidemic of HIV/AIDS, or atomic power self destruction.

5

Homo Sapiens

The Homo Sapien species is the only human species in existence today. All others have become extinct. This is the only species that has the ability to think and to make choices. There is evidence discovered that this species is at least 150,000 years old. On June 11, 2003 ABC—TV News showed a skull of a Homo Sapien found in Ethiopia last year and announced it was dated to be 150,000 years old. This is very young considering that each species normally last an average of four million years. The Neanderthal lasted from around 200,000 years ago until they became extinct around 30,000 years ago. They had a hard life competing for food in a harsh environment, and most of their life was spent during the Ice Age.

"Every single one of the 6 billion people on the planet today is descended from the small group of anatomically modern humans who once lived in eastern Africa." ("Mapping Human History" by Steve Olson.) All of these people have the same number of chromosomes, 46, 23 from the mother and 23 from the father. Each species has its own number of chromosomes, therefore, it would not be possible for our species to successfully mate with a different one. Although, God knows we try. A horse and a donkey will mate and produce a hybrid mule, however the mule cannot reproduce.

Our species consist of a number of different races born in Africa and extending throughout the world. These different races came about by evolution caused by the many environmental conditions. Examples of these races are European, Slavic, Norse, Asian, African, Middle Eastern, Australian, the Americas, and Polynesian, etc. all having the same number of chromosomes Twenty-three pairs. It is not known when God gave man a soul but there is evidence that up until about 50,000 years ago our species only reacted and did not think nor reason. That is why some believe that there was a "Big Mind Bang" or "Mind Change" around 50,000 years ago. This does not bother me because it really doesn't matter when God does what he does and everything fits the evidence. We will discuss this subject in the next chapter.

Homo Sapien is the most successful species on the planet. We may eventually use up all of our natural resources, or, cause our own demise using atomic energy.

6

The Mind Change

No one knows for sure if there is a superior being in control of the universe, however, we can all long to hope that there is life eternal. Many people really believe that this is true, in fact it is what keeps us going. There is nothing known to science that can prove, nor disprove, the existence of an all powerful, loving God. There is nothing to prove that we have a God given soul, but to have a soul and a belief in God with a promise of eternal life, gives us a wonderful peace of mind.

This chapter, "The Mind Change" is written as a story based on known scientific facts as of 2,000 years after the Birth of Jesus Christ. I trust you will read this story with an open mind and decide for yourself if it has meaning for you. If not, it's just another story. Now for the mind change, or the "Mind Bang". About 50,000 years ago our species of modern man, Homo Sapien or "wise man", with an emphasis on *wisdom* began to leave evidence of making jewelry, very fine tools, and other signs that a fully modern man that had evolved. This was a wise man with a brain capacity of 1,350 milliliters, the average capacity for modern man. These were the first signs of man thinking and making choices from a scientific view.

Most of the other prehistoric human species had become extinct. Homo Habilis, who lived about 2-1.5 million years ago; Homo Erectus with all of his different races, Java man, Peking man, and others, who lived 1.6 million to 200,000 years ago. All except the Neanderthal man, with a brain capacity of 1400cc average, who had arrived around 200,000 years ago and would live among the Homo Sapiens for another 20,000 years before becoming extinct some 30,000 years ago. Homo Sapiens dated back, as far as 150,000 years ago, but for the first 100,000 years he was nothing special. He reacted just like all previous humans, or hominids, by instinct. Like all other earthly mammals he could not reason nor make choices.

Then 50,000 years ago that changed and God gave our species a soul! God gave man the gift of knowledge and wisdom and he was in fact a new creation, because he was different. Neanderthal had a proper brain capacity, but he lacked a soul.

For any new species there has to be a first and this first man of our species could have been called, "Adam" and his spouse could have been called, "Eve". That species was first discovered in a valley in Africa, and that place could have been called, "Eden". God's gift of a soul allowed this species to think and to make choices for themselves, and if they followed God's laws, the Ten Commandments, they could be eligible for their soul to have eternal life. Not the body, the body is obeying the laws of evolution and extinction but the soul belongs to God forever, if he so chooses.

This new species of man with a soul began in Africa and in time traveled from Africa through the Middle East, to Asia, then on to North America, to Eastern and Western Europe. They traveled through entire world, known and unknown, even though the people of the Middle East did not know about the rest of the world. One group of theses new species of man was living on the eastern end of the Mediterranean as the great glaciers from the Ice Age were melting into the Mediterranean One of them was named Noah, a man of God. This is the story we all know from the Bible. In his dreams God told Noah that a great flood was coming and he should prepare a boat to save his family and his live stock. (Gilgamesh, in Babylon, had written a thousand years earlier about a flood, very similar to the Noah story.) But none of these stories were written down at the time because writing had not been invented. (Writing began around 5,000 years ago and printing only 900 years ago.) All of these stories were handed down from father to son from the time of the flood until they could record the stories.

There was a great flood that occurred 7,600 years ago, at the end of the Ice Age, when the ice melted into the Mediterranean and flooded into the Black Sea killing many people and animals. Some people survived and it was not a worldwide flood, but to the people involved it seemed as if it were. You can read about this version in a book entitled, "Noah's Flood" written by William Ryan and Walter Pitman.. If you decide not to accept this version of the Genesis story and prefer to take the Bible version literally, that is your choice.

Either version will give you life eternal if you have followed the Ten Command-ments and God grants you Grace as explained by Jesus Christ in the New Testa-ment This and other versions and interpretations of the Bible are for you to study and decide for yourself. The rest of the Hebrew history in the Bible proceeded on to Abraham. Abraham begot the Hebrew tribes through Isaac and the Arab tribes through Ishmael. Moses delivered the Hebrews from the Egyptians and brought down the Ten Commandments that everyone should live by.

7

The Periodic Table

Although the world is varied and complex, everything in it, i.e. air, water, rocks, and living tissue, and an infinite number of other objects and materials, which are all around us, it is actually made up of only a limited number of chemical elements. We know today that only 91 such elements exist naturally on the Earth. They range from hydrogen, the lightest element, to uranium, the heaviest. Actually, several more elements do exist, but these have to be made artificially in chemical laboratories.

The basic components of each chemical element are atoms. The atoms of an element consist of three kinds of particles: protons, neutrons, and electrons. Protons and neutrons exist at the core, or nucleus of the atom. One of the important ways in which these two kinds of particles differ from each other is that each proton carries a single, positive electric charge, whereas a neutron carries no electrical charge. The electrons are much smaller than either the protons or the neutrons and each carry a single negative electric charge. The electrons orbit the nucleus at a distance in complex paths. Under normal circumstances, the number of electrons orbiting the nucleus of a particular atom is equal to the number of protons in the nucleus of that atom. This means that the over all balance of the electrical charge from the positive protons and the negative electrons orbiting the nucleus is equal.

The key event that led to the modern understanding of the atom was the discovery that atoms are made up of electrons, protons, and neutrons. Thus, despite its name, which derives from the Greek word for "indivisible," the atom could be divided into smaller components. Until 1897 the atom was alone, the smallest unit of matter.

In April 1897, Joseph Thompson, a professor of physics and director of the prestigious Cavendish Laboratory at Cambridge University in England, announced the discovery of the electron. Then Ernest Rutherford, a New Zealand physicist who had been a pupil of Thompson's and who was a professor at Cambridge University, moved to the next step for the understanding of the atom. Rutherford discovered that the atom had a nucleus and that one of the particles in the nucleus was the positively charged proton. Finally, in 1932, an English physicist Sir James Chadwick, who also worked at the Cavendish Laboratory in Cambridge, discovered that yet another particle existed in the nucleus of atoms. This particle was called the neutron and had a mass almost the same as the proton, but it had no electric charge.

These discoveries along with the work of a young physicist, Henry Mosley, led to the reason for Mendeleev's success with his periodic table. Mosley discovered that every element had a different number of protons in the nucleus. The number of protons came to be called, the "*atomic number*" designated by the letter Z, and is always a whole number in the elements of the Periodic Table. (The atomic weight, in the Periodic Table, is a measure of total number of protons and neutrons in the nucleus.)

The results are that atoms are normally electrically neutral with an equal number of electrons and protons. This means, for example, in the Periodic Table, that oxygen, with an atomic number of 8, has eight protons in its nucleus and eight electrons orbiting around the outside. However, the atom can have a different numbers of neutrons.

PERIODIC TABLE OF THE ELEMENTS

Atom #	Symbol	Element	Type Gas	Atom #	Symbol	Element	Type Gas
1	H	Hydrogen		57	La	Lantha-num	Noble
2	He	Helium	Noble	58	Ce	Cerium	
3	Li	Lithium		59	Pr	Praseod ymium	
4	Be	Beryllium		60	Nd	Neody-mium	
5	B	Boron		61	Pm	Prome-thium	
6	C	Carbon		62	Sm	Samar-ium	
7	N	Nitrogen		63	Eu	Europiu m	
8	O	Oxygen		64	Gd	Gadolin-ium	
9	F	Fluorine		65	Tb	Terbium	
10	Ne	Neon	Noble	66	Dy	Dyspro-sium	
11	Na	Sodium		67	Ho	Holmium	
12	Mg	Magnesium		68	Er	Erbium	
13	Al	Aluminum		69	Tm	Thulium	
14	Si	Silicon		70	Yb	Ytter-bium	
15	P	Phospho-rus		71	Lu	Lutetium	
16	S	Sulfur		72	Hf	Hafnium	
17	Cl	Chlorine		73	Ta	Tanta-lum	

Atom #	Symbol	Element	Type Gas	Atom #	Symbol	Element	Type Gas
18	Ar	Argon	Noble	74	W	Tung-sten	
19	K	Potassium		75	Re	Rhe-nium	
20	Ca	Calcium		76	Os	Osmium	
21	Sc	Scandium		77	Ir	Iridium	
22	Ti	Titanium		78	Pt	Platinum	
23	V	Vanadium		79	Au	Gold	
24	Cr	Chromium		80	Hg	Mercury	
25	Mn	Manganese		81	Ti	Thallium	
26	Fe	Iron		82	Pb	Lead	
27	Co	Cobalt		83	Bi	Bismuth	
28	Ni	Nickel		84	Po	Polo-nium	
29	Cu	Copper		85	At	Astatine	
30	Zn	Zinc		86	Rn	Radon	Noble
31	Ga	Gallium		87	Fr	Fran-cium	
32	Ge	Germa-nium		88	Ra	Radium	
33	As	Arsenic		89	Ac	Actinium	
34	Se	Selenium		90	Th	Thorium	
35	Br	Bromine		91	Pa	Protac-tinium	
36	Kr	Kripton	Noble	92	U	Uranium	
37	Rb	Rubidium		93	Np	Nep-tunium	
38	Sr	Strontium		94	Pu	Pluto-nium	

Atom #	Symbol	Element	Type Gas	Atom #	Symbol	Element	Type Gas
39	Y	Yttrium		95	Am	Americium	
40	Zr	Zirconium		96	Cm	Curium	
41	Nb	Niobium		97	Bk	Berkelium	
42	Mo	Molybdenum		98	Cf	Californium	
43	Tc	Technetium		99	Es	Einsteinium	
44	Ru	Ruthenium		100	Fm	Fermium	
45	Rh	Rhodium		101	Md	Mendelevium	
46	Pd	Palladium		102	No	Nobelium	
47	Ag	Silver		103	Lr	Lawrencium	
48	Cd	Cadmium		104	Rf	Rutherfordium	
49	In	Indium		105	Db	Dubnium	
50	Sn	Tin		106	Sg	Seaborgium	
51	Sb	Antimony		107	Bh	Bohrium	
52	Te	Tellurium		108	Hs	Hassium	
53	I	Iodine		109	Mt	Meitnerium	
54	Xe	Xenon	Noble	110	Uun	Ununnilium	
55	Cs	Cesium		111	Uuu	Unununium	

Atom #	Symbol	Element	Type Gas	Atom #	Symbol	Element	Type Gas
56	Ba	Barium		112	Uub	Ununbii-num	

8

Evolution—Carbon Dating

✦

Radioactive Isotope Dating System

Isotopes are atoms of the same elements but have different atomic weights. Radioactive isotopes, or radioisotopes, of elements such as carbon have wide uses as tracers by scientist looking for a way to measure age of samples. Geiger counters can follow their motion.

More than 270 radioactive isotopes occur in nature. About 50 heavier isotopes, including those of uranium and radium, occur naturally but are radioactive and give off particles. Radioactive atoms decay, or lose particles, and change into lighter weight isotopes of the same element. They belong to three radioactive decay series, which start with U^{238},

U^{235}, and Thorium-232. These heavy atoms decay into various isotopes until they eventually become stable isotopes of lead. The rate at which radioactive isotope decay is measured by a term called half-life, or the time required for half the atom in a sample to decay. *Every isotope has a specific half-life*. There are isotopes in the radioactive series that decay more slowly. The radium isotope Ra^{226} has a half-life of 1,600 years. Some may range from 1,600 to billions of years. A radio-isotope of carbon 14 has been used to date materials that are older than written history. Geologists use other radioisotopes to date rocks and fossils using Geiger counters.

A Geiger counter picks up and counts the rays given off by radiocarbon. By this means, scientist can determine the age of once-living material up to about 44,000 years. A man gives off 918 disintegration rays per hour per gram of radiocarbon. After death, the radiocarbon gives off rays to a known rate. It loses half its radio-activity in the first 5,600 years and half the remainder each 5,600 years that follow.

Radiocarbon or carbon 14 is a radioactive isotope of carbon. It can be used for finding how old things are by means of radiocarbon dating of ancient items. Over the years scientists have gained much knowledge about prehistoric man, prehistoric animals, and changes in the earth's climate. Radiocarbon is formed by cosmic rays. When atomic particles reach the earth's atmosphere, they smash different kinds of atoms in the air, breaking them down into neutrons, protons, mesons, and other particles. Some of the neutrons strike atoms of nitrogen. This causes the nucleus of the nitrogen atom to disintegrate and give off a proton. The atom then becomes radiocarbon, which has an atomic weight of 14. Radiocarbon is heavier than ordinary carbon, which has an atomic weight of 12.

When a tree is cut down, it stops taking in radiocarbon. But the radiocarbon already in the wood continues decaying at a constant rate. If wood from a certain tree was used to make a casket for an ancient noble. We could measure the radiation from the radiocarbon left in the wood, and learn how old the wood is. This could tell us when the person died. In dating an object by its radiocarbon contents the scientist first burns a sample of it to convert it to carbon dioxide. The carbon dioxide is purified and reduced to pure carbon. After this reduction, the amount of radioactivity remaining in the carbon is measured with a Geiger counter. This is to give you a better idea of how scientist can find the age of various materials that were once alive on the earth. To understand this process is part of our gaining knowledge in preparation for the chapter, "You are what you know."

9

Genome—DNA—RNA—Mitoch ondria DNA

DNA is a thin chainlike molecule found in every living cell on earth. The structure of this molecule is a double helix. The discovery of this structure in 1953 by James Watson, Francis Crick, and Maurice Wilkins earned them a Nobel Prize in 1962. Also, involved with and playing an important role, in the discovery of the structure of DNA in 1953 was Rosalind Franklin, a pioneer molecular biologist. Franklin was responsible for the x-ray photos of DNA molecules, which led to the understanding of the structure of DNA. Unfortunately, she died at the early age of 37, in the year 1958 of ovarian cancer, four years before the others received the Nobel Prize.

DNA directs the formation, growth, and reproduction of cells and organisms. Short sections of DNA called genes determine heredity—that is, passing on of characteristics—in living things. DNA is found mainly within a cell's nucleus, in threadlike structures called chromosomes. Chromosomes carry the genes that convey hereditary characteristics and they are constant in number for each species. The human genome has 23 pairs of chromosomes for the human species.

"The human genome—the complete set of human genes—comes packaged in twenty-two pairs that are numbered in approximate order of size, from the largest (number 1) to the smallest (number 22), while the remaining pair consists of the sex chromosomes: two large X chromosomes in women, one X and one Y in men. In size the X comes between chromosome 7 and 8, where as the Y is the smallest."[1]

Genomes are written entirely in three-letter words, using only four letters: A, C, G, and T.

---

1.   "Genome", by Matt Ridley

A = adenine   C = cytosine

T = thymine   G = guanine

DNA = deoxyribonucleic acid
RNA = ribonucleic acid

The start of human life takes place when the egg from the mother is fertilized with the sperm from the father. The egg cell has a tiny nucleus that contains thread like objects called chromosomes. Inside of the chromosomes is the genes that will direct your growth, determine your sex, and will assist in deciding what you will grow up to be. The sperm from your father also contains chromosomes. When the egg and sperm get together their chromosomes unite in the nucleus. Nothing happens until the two sets of chromosomes get together, pair up to make 23 pairs for a total of 46 chromosomes. One-half of the chromosomes come from your mother and one-half from your father. Then the egg begins to divide, the chromosomes copy themselves so that each new cell has 23 pairs of chromosomes. Voilà! To determine sex, the mother has two X-chromosomes and the father has one X and one Y chromosome. If the sperm has an X-chromosome it is a girl and if the sperm has a Y-chromosome it is a boy. The father decides!

Before a cell divides, it duplicates its DNA. The two chains of polynucleotides separate lengthwise, splitting the bonds between the base pairs, (the double helix structure which looks like a spiral ladder). The separated chains, each resembling half a ladder split down the middle, serve as templates (patterns) for two new DNA molecules. Therefore, when the cell divides each of the resulting cells receives identical DNA molecules.

RNA is a complex molecule that plays a major role in all living cells. RNA molecules help produces substances called **proteins**. Proteins are chains of smaller organic molecules known as **amino acids**. The body uses proteins to build cells and to carry out the cells' work. RNA is similar in structure to DNA and like DNA all RNA molecules contain hundreds of smaller chemical units called nucleotides.

Different types of RNA perform different jobs. One type, known as messenger RNA or mRNA, copies chemical instructions from DNA for making proteins.

The mRNA then leaves the nucleus and carries the instructions to protein-making cell structures, called **ribosome**. These instructions tell the cell how to put amino acids together in the correct order to make a specific protein.

In science DNA is used for fingerprinting, also known as DNA typing. The cells may come from almost any body fluid or tissue, including bone, blood, semen, hair, or teeth. In order to obtain DNA from very old bones like those in the "Iceman" found in the Italian Alps on Thursday 19 September 1991, and later determined to have been between 5,000 and 5,350 years old. To get the proper amounts of DNA the scientist had to decide which gene to amplify because they knew there wasn't going to be much, if any, DNA left after such a long period of time. They decided to maximize their chances by choosing something called **mitochondrial** DNA.

"Mitochondria are tiny structures that exist within every human cell. They are not in the cell nucleus, the tiny bag in the middle of the cell that contains the chromosomes, but outside it in what is called the **cytoplasm**. Their job is to help cells use oxygen to produce energy. The more vigorous the cell, the more energy it needs and so the more mitochondria it contains. Cells from active tissues like muscle, nerve and brain contain up to one thousand mitochondria each.

Mitochondria is unique version of DNA and not like the DNA found in the nucleus, which is inherited from both parents, everyone gets their mitochondria from only one parent—their mother. The cytoplasm of a human egg cell is stuffed with a quarter of a million mitochondria. In comparison, the sperm have very few mitochondria, just enough to provide the energy for swimming up the uterus as they home in on the egg. After the successful sperm enters the egg to deliver its package of nuclear chromosomes it has no further use for the mitochondria, and they are jettisoned along with the tail. Only the sperm-head with its package of nuclear DNA enters the egg. The plump, fertilized egg now has nuclear DNA from both parents, but its only mitochondria are the ones that were in the cytoplasm all along—and they all come from the mother. For the simple reason, mitochondrial DNA is always maternally inherited." [2]

---

2.   "The Seven Daughters of Eve" by Bryan Sykes

10

Chronology

✦

{Relating to my essay}

This chronology and the geological periods listed in this essay are my own estimates based on the many list available in as many different encyclopedias and other reference books that have been published in the twentieth century. They all vary as the search for knowledge progresses into the twenty-first century, where the means for determining such information, improves its technique and reliability for accuracy.

I will use the term "Years Ago" in place of "BC" (Before the Birth of Christ), since most Biblical scholars now place the birth of Christ as the year 5 BC. After year 1 BC, I will begin to use our current calendar that we use today. If you have reason to compare the dates given with the term "BC" you may subtract 2,000 years from the years ago column to ascertain the BC date. The "Years Ago" date is the time between a ten year period of 1995 to 2005. After that you will have to make the adjustments you deem fit.

Years Ago	Event
65,000,000	Dinosaurs become extinct.
57,000,000	First horse & first monkeys.
34,000,000	First of the great apes. (Chimpanzee, gorilla, and baboon).
5,000,000	First of the hominids. (Any two-legged primates.)
500,000	Homo erectus. (Prehistoric races now extinct.)
200,000	Neanderthal. No DNA information at this time. (Now extinct.)

150,000	Homo Sapien. Our species with many different races.
120,000	Same temperature and ocean water levels as today.
100,000	Ice Age began.
50,000	The Big Mind Bang. God gave our species a soul with the power to think and to make choices. This was done in the garden of Eden in Africa. Adam and Eve were the first of our DNA to have a soul.
45,000	Ursula, the name given by Bryan Sykes to the first of his seven daughters of Eve. Ancient bones of female Homo Sapien DNA during Ice Age. The Neanderthal species were still active.
38,000	Very few Neanderthal remaining, no DNA found.
37,000	Ancient counting was limited to tally sticks. First known was a baboon fibula with 29 notches found in Africa.
32,000	Wolf bone with 55 notches found in Morava, Chech Republic.
25,000 25,000	Xenia, the name given by Bryan Sykes to the second of his seven daughters of Eve. Bones found in the great plain from lowland Britain in the west to Kazakhstan in the east, before the English Channel existed. She was born during the Ice Age. The Neanderthal are now extinct.
20,000	Ice Age melt down has started, but is still very severe. Helena, the name given by Bryan Sykes to the third of his seven daughters of Eve, born in Europe, close to the Mediterranean.
17,000	Velda and Tara, names given by Bryan Sykes to the fourth and fifth of his seven daughters of Eve. Velda was born in Spain and Tara was born in Italy, both were born in the Ice Age.
15,000	Katrine, a name given by Bryan Sykes to the sixth of his seven daughters of Eve, born near Venice Italy.
12,500	Return of Ice Age for about 1,000 years. (Younger Dryas).
11,400	Ice Age ends.
10,000	Jasmine, a name given by Bryan Sykes to the last of his seven daughters of Eve, born in Syria, about one mile from the River Euphrates. The Ice Age was at an end. Water was now in the English Channel.
7,600	The Great Flood, which occurred at the end of the Ice Age as the ice, melted into the Mediterranean and over flowed into the Black Sea. (See the book, "Noah's Flood" by William Ryan and Walter Pitman.)

6,004	This is the date calculated by Archbishop James Usher (1581-1656), using the Book of Genesis as a reference, as the date God created the universe. This was used by the Church to combat the works of Copernicus and Galileo.
5,800	This is the date given by Gilgamesh in Babylon Myth for the Great Flood. Date established from Gilgamesh Epic.
5,000	Writing was invented in Mesopotamia, the land between the Tigris and Euphrates rivers. Astronomy used in Egypt.
4,800	Approximate time of Noah's flood in the Bible.
4,000	The beginning of Abraham in Bible history. Abacus used for counting. In China and Babylon they are practicing astronomy.
3,500	The Exodus of Moses and the Ten Commandments.
3,460	The conquest of Canaan in the Bible.
3,102	The Book of Judges in the Bible.
2,982	Samuel, the last and greatest of the Judges anointed Saul to be King. The three kings, Saul, David, and Solomon, reigned, each of them about 40 years. This was Israel's "golden age."
2,722	The two kingdoms, Israel in the north and Judah in the south.
2,587	Israel was in captivity for 135 years in Babylon.
2,538	Cyrus allowed the people of Israel to return to Jerusalem. The temple and wall were rebuilt under Zerubbabel and Ezra.
2,400	Democritus believed that all matter was made of atoms.
2,391	The Old Testament closes.
2,384	Aristotle born and was the tutor of Alexander the Great.
2,331	The Jews were under Persian rule until Alexander the Great.
2,325	Euclid's elements, a point, a line, a surface, and a circle.
2,290	Archimedes—Lever, pulley, screw, etc.
2,275	Eratosthenes, Geographer, astronomer, and mathematician calculated the circumference of the Earth.
2,190	Hipparchus astronomer calculated the length of a year 365 days, 5 hours, and 55 minutes, the month 29 days, 44 minutes, and 2.5 seconds. He was short by 1 second. He also adopted a 360-degree circle and discovered movement of equinoxes.

2,167	The Jews were under the control of Greek kings.
2,100	Hindu mathematicians invented the concept of zero. Arabs picked up this concept 900 years later.
2,063	General Pompey of Rome conquered Jerusalem.
2,046	The Julian calendar is introduced by Julius Caesar in which the ordinary year had 365 days and the months were the same as in the Gregorian, or New Style, calendar now used.
2,037	Herod the Great was king under the Romans.
2,005	The Birth of Jesus the beginning of Christianity.
AD Began	The concept of a zero was not known at that time, so the first 100 years began with the year 1 instead of the year zero. Therefore, with our calendar, known as the Gregorian Calendar and introduced by Pope Gregory in the year 1582, the years 0 to 99 is known as the first century, the years 100 to 200 is the second century and so on.
28	Jesus is put to death.
37–40	Conversion of Saul to Paul on the road to Damascus.
90–168	Claudius Ptolemy-Earth-centered Universe as stated by Aristotle and was taken seriously for about 1,400 years until Tycho Brahe and Galileo.
598–665	Brahmagupta The concept of zero along with the decimal system came back and was picked up by Arabs and Arabic numbers about 800 AD and taken to Spain with the Moors.
1100 1100	The universities in Moorish Spain introduced the concept of the zero and the decimal system to the merchants of the Mediterranean area and thus to all of Europe in time for the printing press in 1440.
1170–1240	Fibonacci invents Algebra in 1202 AD using Arabic numerals 0-9 and adopted by Muslims.
1290	Fall of the Roman Empire—Start of Middle Ages with very little learning-To the start of the Renaissance (New Birth) in the late 1200's to 1500's.
1440	Johannes Gutenberg invented the printing press.
1473–1543	Copernicus—Sun-centered Universe and the Reformation.
1492	Columbus discovers America
1546–1601	Tycho Brahe—Greatest astronomer observer ever. Supplied Kepler with information to make his Laws of Planetary Motions.

1550–1617	John Napier in 1614 invented Logarithms.
1564–1642	Galileo Galilei using new telescope (Invented in Holland) in 1610 had a book on stars. Proving Copernicus was right. Did great work in study of motion.
1565	Fossil Objects discovered.
1571–1630	Johannes Kepler in 1609 introduced his Laws of Planetary Motion.
1600	Natural Magnetism
1629–1695	Christian Huygens developed a wave theory of light.
1642–1727	Isaac Newton—Laws of Motion—Principia—Calculus—Gravity. Was born the year Galileo died.
1700–1782	Daniel Bernoulli—Kinetic theory of gases.
1728–1777	Johann Heinrich Lambert—Pi the constant was known to earliest civilizations. The Greek letter π to denote the periphery of a circle did not appear until 1768.
1745–1827	Alassandro Volta (1745-1827) Invented the electric battery in 1799.
1766–1844	John Dalton introduced his atomic theory of matter in 1803.
1776	United States Independence—July 4, 1776.
1791–1867	Michael Faraday Discovered the principle of electromagnetic induction in 1831.
1791–1872	Samuel F. B. Morse developed the telegraph and code in 1844.
1804–1892	Invention of the term *dinosaur* by Richard Owen in 1842.
1807–1873	Ice Ages discovered by Jean Louis Agassiz in 1840.
1809–1882 1809–1882	Charles Robert Darwin (1809-1882) published his famous book on evolution, "Origin of the Species" in 1859.
1818–1889	James P. Joule—The unit of measure of energy, the joule, named in his honor for his work in heat and energy.
1822–1884	Gregor Mendel discovered the Laws of Inheritance in 1865.
1831-1879	James Clerk Maxwell—Electromagnetic theory of light.
1845–1923	Wilhelm K. Roentgen discovered X-rays.
1846	Discovery of the planet Neptune.

1852–1908	Antoine Henri Becquerel along with Pierre and Marie Curie discovered radioactivity in 1896,
1852–1931	Albert Michelson determined the exact speed of light.
1856–1940	Joseph John Thomson discovered the electron.
1857–1894	Heinrich Rudolph Hertz opened the way for the development of radio, television and radar with his discovery of electromagnetic waves, between 1885-1888, which were predicted by James Clerk Maxwell.
1858–1947	Max Planck—The Quantum theory.
1861–1865	United States War Between the States.
1879–1955	Albert Einstein discovered the Theory of Relativity in 1905.
1885–1962	Niels Bohr—Atomic structure and radiation.
1887–1961	Erwin Schrödinger—Quantum Mechanics.
1891–1974	Sir James Chadwick—Discovery of the neutron.
1901	Blood Groups discovered. Marconi invented wireless radio.
1903	The Wright brother's first airplane flight at Kitty Hawk, NC in December 1903. Then they had to learn the most important part of how to actually control the flight, the take off, the landing, turning and the many different flight problems.
1910	Discovery that genes are inheritance.
1929	Edwin Hubble—The expanding Universe. The Big Bang.
1942	Stephen Hawkins was born in 1942 on the anniversary of Galileo's death. Wrote the popular books, "A Brief History Of Time" and "The Universe in a Nut Shell".
1953	Discovery of the structure of DNA as a double helix by James Watson, Francis Crick, and Maurice Wilkins. Nobel Prize in 1962. Rosaline Franklin's work in x-ray with DNA was essential to the discovery. Unfortunately she died in 1958 of ovarian cancer, four years before Nobel prize award.
2001	"The Seven Daughters of Eve" by Bryan Sykes. (DNA)
2002	"Mapping Human History" by Steve Olson. (DNA)

11

You Are What You Believe

First and foremost I know that I believe in God and I am a Christian. I believe God created the universe and the earth, but not at the same time. I believe that God gave man a soul, which includes the ability to think, to reason, and to make choices. I believe in the teachings of Jesus Christ. I also believe in evolution as explained by Charles Darwin. I believe evolution and religion are compatible. Evolution concerns the body and religion concerns the soul. God created the earth and all life, therefore He is in charge of the maintenance of life and changes to allow life to adapt to its environment. That maintenance is called evolution. Charles Darwin never explained, to my satisfaction, how each new species began if it had a different DNA and a different set of chromosomes? Darwin did not know about DNA and chromosomes. He lacked the knowledge that we now have, even so, DNA has proven that Darwin was correct about the changes evolution makes in all species to allow them to adapt to environmental conditions.

In the PBS-TV series, "Evolution—a journey into where we are from and where we are going" written by Carl Zimmer and presented on PBS in January 2002. There is a companion book by Carl Zimmer and I purchased both and recommend both the VCR tapes and the book.

There is a conflict among young people today, the same as it was fifty years ago, regarding the things they encounter when they leave home to begin their education in colleges and universities. They are introduced to new thoughts about the universe and morality. They discover differences in the beliefs they were taught, at home and in their religious environment, and what scientist now believes. These thoughts cause great concerns among the students and their parents. The young people are my main reason for writing this essay; it is not an either-or question, between creation and evolution, they are one and the same. God is involved with both creation and evolution of each species.

Opponents of evolution may say, "What the Bible says is what it means. If it says God created the Earth in a day. A day is a day, it is not some other period of time, it is a day". The first Bible was not written in English. The Old Testament was written in Hebrew and the New Testament was written in Greek. Translations of the Hebrew and Greek into other languages could have meant something other than a day. Fundamentalist also say, "When someone says to you, 'Millions of years ago', say to him, 'Were you there?'" Reply, "I don't have to be there, God gave me the power to reason." Fundamentalist also say, about geologist finding animal bones in the mud, "It is logical that they found dead animals in the mud because all living things, except those in the Ark, died in the ***great flood*** that covered the earth." He is what he believes, and if that is what he believes, then that is his choice and is his right to believe whatever he wants. It is my right to believe what I want to believe.

The entire world, at the time of the great flood, was the area around one end of the Mediterranean Sea. Gilgamesh, was a man from a story in Babylon mythology that tells a tale of a great flood, just like the Noah story, that occurred many years before, but it was written 1,000 years before the Hebrew account. In a recent PBS-TV NOVA account of the great flood, based on the Book entitled, "Noah's Flood", written by William B.F. Ryan and Walter Pitman. They say that the actual "great flood" occurred when the salt waters of the Mediterranean Sea flooded into the Black Sea, at the end of the last great Ice Age, 7,600 years ago. I urge you to read this book and purchase the NOVA PBS series account.

The great flood in this story was 2,600 years before the invention of writing. Therefore, the telling of this great flood story was handed down by word of mouth by the different peoples of this vast area giving the only explanation they could come up with. So you have a Babylonian story and a Hebrew story, almost identical, written 1,000 years apart. You can believe, or not believe, that global melting at the end of the Ice Age caused the great flood. There are those who do not believe in the Ice Age ever happened. That is their right. Whichever way it happened, it should not affect your belief in God. The only thing that really matters is how you live your life and how you obey God's laws.

I accept the Old Testament of the Bible as a history of the Hebrew people and their relationship with God and all of the other peoples that they come in contact with during the Old Testament time period. This period goes back approximately three to four thousand years BC. (Archbishop Ussher, during the Middle Ages, estimated the age of the earth, according to the Book of Genesis, as beginning on October 23, 4004 BC.) Some Bible fundamentalist still believes this to be true, and I believe they have the right to believe whatever they wish. I only wish they would let me believe whatever I want to believe. The New Testament is the story of the life of Jesus Christ and the beginning of the early Christian church. The New Testament was first written in Greek. I will leave the discussions of the Bible to the religious scholars.

The people writing this history of the Bible did so with the knowledge that was available to them at the time it was written. That time goes back to the very beginning of **recorded** history, 5,000 years ago, (3,000 BC when writing was invented) and would include stories told in earlier times about earlier events, things that were handed down through time from father to son. That was all written by hand, and copies were made by hand. (Printing was not invented until the 15th century AD) Later it was translated and again copied by hand. There were very few people who could read and write. All of this copying and re-copying; translating and re-translating, by people of marginal ability, was bound to cause many human errors. A good reference to understand these problems can be found in the book, "In the Beginning" by Alister McGrath. The story of the King James Bible and how it changed a nation, a language, and a culture. ("In the 16th century, to attempt to translate the Bible into a common tongue wasn't just difficult, it was dangerous.")

The three main things I use from the Old Testament are the Ten Commandments, the Twenty-third Psalm, and the Book of Proverbs. I think the Old Testament is a great literary gem with many wonderful stories, but I do not take the Old Testament literally; and do not consider myself less a believer in God for not doing so. You are what you believe. Proverbs is important because it teaches that God wants us to continue to learn and to grow and to gain wisdom.

I remember listening to the radio back in the late 1940s, before television, to a Sunday night program featuring Bishop Fulton J. Sheen. He was talking about God in the Old Testament and he said that it was like a little boy who s mother caught him in the cookie jar. The little boy knew he was doing wrong, so he said, "Now mommy don't be mad with me." He expected punishment because he knew he was doing wrong. The Hebrews knew when they had sinned and they expected God to punish them. They thought he was a God of wrath because that is the way they would have reacted. Bishop Sheen said that the Hebrews did not know that God was a God of love as taught by Jesus.

Hebrews wrote the first Bible in the Hebrew language. Writing was invented by the Egyptians about 3000 BC, followed in other Mesopotamian countries soon after. So the spoken language must have handed down the Bible until there was a Hebrew written language invented. There would have been misinterpretations when translating Hebrew into other languages. (Mark Twain says, "Don't read about health cures from a book as you could die from a misprint.)

Another problem for taking the Bible literally is with numbers. At the time the Bible was written, both Old and New Testaments, there was no concept of the "zero" and no known use of the decimal place system. These two facts would have caused problems with very large numbers. People could count long before they could write. (If you look at the Chronology, in the previous chapter, you will note that 37,000 years ago ancient counting was limited to tally sticks. First known was a baboon fibula with 29 notches was found in Africa.) This was a crude way of counting, but served the ancients well. They had words for each amount. It is also known that the early Babylonians and Egyptians had a verbal word that meant "plenty" for large numbers. When writing was invented they would have had a written word for plenty. Each person would have to figure out how much a "plenty" was. If you were one of the scribes copying a Bible and the reader said, "Now Solomon had plenty of wives and concubines." You could have interpreted the amount to suit your self.

Roman numerals did not come into being until around 200 BC, and, even with its long life, Roman numerals never understood the concept of "zero". Roman numerals were used until the adoption of Arabic numbers came into vogue. They are still used today for certain artistic uses, but seldom.

Arabic Numbers

Scholars are not sure how Arabic numerals originated. But the symbols (1,2,3,4,5,6,7,8 & 9) for all the digits except zero most likely originated with the Hindus in India, as early as the 200's BC. The concept of a zero and using a place, or decimal point, was introduced by the Hindus, around 100 BC. The Arabs borrowed this knowledge around 800 AD adopted the concept of zero using the symbol "0", and the decimal, or place; system. They brought these to Spain about 900 AD. Commercial traders traveling in the Mediterranean area carried them to the rest of Europe about 1100 AD and by scholars while attending the universities in Spain. The Arabic number symbols, and places system, along with the zero came into general use in Europe when the symbols were standardized. This was brought about by the invention of the printing press during the middle 1400's AD. You are what you believe.

Calendars

The Hebrew calendar, according to tradition, was supposed to have started with the creation at a moment 3,760 years and 3 months before the Christian era. You must add 3,760 years to the present calendar to get the number for the Jewish calendar.

Our calendar came down from Early Romans who twice gave the early world a calendar. There are about 40 traditional or religious calendars in use around the world, such as, Jewish, Islamic, Hindu, and Chinese. However, it is the calendar of Julius Caesar as slightly modified by Pope Gregory XIII that is now used as the world norm. This calendar has a solar year which last 365 days, 5 hours, and 49 minutes.

Our calendar is supposed to be based on the year Jesus Christ was born. All dates listed before that year are listed as BC, or Before Christ. Dates after that year are listed as AD, or Anno Domini, meaning "in the year of our Lord." Non-Christians often write BCE, for "before the Christian era", or just CE, for Christian era." (It is estimated that Jesus was born 4 to 6 years later.) It should have no effect on your religion or how you live. The first century AD was supposed to have been known as 0 AD, but the concept of a zero was not known at that time, so it started as the year 1 AD. For that reason our 20th Century was called the 19th Century and we are presently in the 21st Century. To further confuse you, in this essay the chapter on the Chronology, starts with "Years Ago" instead of BC. And since the birth of Christ it lists the years ago as AD. So years ago, which represents BC, can actually mean some time between that number and 4 to 6 years later depending on when Christ was born. Or is it earlier? Don't lose sleep over 4 to 6 years, just obey the Ten Commandments because you are what you believe.

12

Wisdom Brings Happiness

✦

(This is a Sunday school lesson for The Luther Snyder Bible Class.)

The purpose of the lesson is to acquire an understanding of wisdom as God's valuable gift. The Scripture for this lesson is from the Book of Proverbs, chapters three and four.

Reading from the Holy Bible, New Living Translation:

Chapter 3:13-18

"Happy is the person who finds wisdom and gains understanding. For the profit of wisdom is better than silver, and her wages are better than gold. Wisdom is more precious than rubies; nothing you desire can compare with her. She offers you life in her right hand, and riches and honor in her left. She will guide you down delightful paths; all her ways are satisfying. Wisdom is a tree of life to those who embrace her; happy are those who hold her tightly."

Chapter 4:1-9

"My children listen to me. Listen to your father's instruction. Pay attention and grow wise, for I am giving you good guidance. Don't turn away from my teaching. For I, too, was once my father's son, tenderly loved by my mother as an only child.

"My father told me, 'Take my words to heart. Follow my instructions and you will live. Learn to be wise, and develop good judgement. Don't forget or turn away from my words. Don't turn your back on wisdom, for she will protect you. Love her, and she will guard you. Getting wisdom is the most important thing you can do! And whatever else you do, get good judgement. If you prize wisdom, she will exalt you. Embrace her and she will honor you. She will place a lovely wreath on your head; she will present you with a beautiful crown.'"

Did you notice during the reading of the scripture from Proverbs that the word "wisdom" was referred to as "she" or "her"? It is a feminine word, God is telling us something! Remember this, as I will bring this up again before the end of the lesson.

The purpose of this lesson is to acquire an understanding of wisdom as God's valuable gift. Knowledge is often confused with wisdom. I will now explain the difference between knowledge and wisdom. When I was in college learning math, physics, and general science, I had the knowledge to know how to tell the age of different materials and mammals using isotopes from various elements. And, now I have the wisdom not to try and find out the age of women.

There is more knowledge available today than at any time in our past and it is increasing at a rate much faster than ever before. We have the opportunity to learn, and hopefully understand, more than we ever knew before. To learn is to change. We change from what we use to believe, and it is sometimes difficult to understand, however, do not be afraid to learn new ideas and beliefs. God gave us a valuable gift in the power to think and to make choices when he gave our species a soul! So it is good to seek new knowledge about evolution, DNA, extinction, the environment, etc. as long as we do not forget the wisdom that God gave us as a basis for our Christian faith and our belief in God.

The wisdom God gave us with The Ten Commandments; with words like the ones in the 23rd Psalm, with the teachings of Jesus Christ, including the Lord's Prayer, the Sermon on the Mount, and the power of Grace. Reading the Psalms, Proverbs and especially the New Testament, is God's gift of wisdom that will make us happy. "Happy are those who find wisdom, and those who get understanding." God also asks you in the scripture to get insight. Insight comes from knowledge and knowledge comes from learning; never stop learning.

As the scripture says, "Let your heart hold fast my words; keep my commandments and live." Remember the Ten Commandments:

1. There is only one God.

2. Do not make idols of any kind nor worship any other gods.

3. Do not misuse the name of the Lord.

4.   Remember to observe the Sabbath day by keeping it holy.

5.   Honor your father and mother.

6.   Do not murder.

7.   Do not commit adultery.

8.   Do not steal.

9.   Do not testify falsely against your neighbor.

10.   Do not covet anything belonging to any person.

Very few people can recite Bible verses anymore. It is a good habit to start.

God created the universe, the earth, and all living things. How He did it, and when He did it, and how long it took to do it, really doesn't matter. God will sort out all the rest. May God grant us the wisdom to see.

BOOK REPORT[1]

Now let me introduce you to a fascinating book, which I have no reason to doubt, and which does not go against my belief. It is entitled, "**The Seven Daughters of Eve**" by **Bryan Sykes**, Professor of Human Genetics, at the University of Oxford, England. (www.oxfordancestors.com). This year marks the fiftieth anniversary of the discovery of the structure of DNA.

"Each of us carries a message from our ancestors in every cell of our body. It is our DNA, the genetic material that is handed down from generation to generation. Within the DNA are written not only our histories as individuals but also the whole history of the human race. With the aid of recent advances in genetic technology, this history is now being revealed. We are at last able to begin to decipher the messages from the past. Our DNA does not fade like an ancient parchment; it does not rust in the ground like a sword of a warrior long dead. It is not eroded by the wind or rain, nor reduced to ruin by fire and earthquake. It is the traveler from an antique land who lives within us all."

Dr. Sykes continues saying, "I have found DNA in skeletons thousands of years old and seen exactly the same genes in my own friends. And I have discovered that to my astonishment we are all connected through our mothers to only a handful of women living tens of thousands of years ago."
This is the basis of his book, "The Seven Daughters of Eve". Bryan Sykes' team is well known in genetic circles. They have been gathering DNA samples for over a decade from both ancient bones to modern day living people, mainly in Europe. These seven women are all from the European area, Since this book was written in 2001, they have now expanded the list of ancient women from seven to twenty-six worldwide, so far, to include women from Africa, Asia, and the rest of the world.

I now need to explain mitochondrial DNA (mtDNA), so please bear with me for five minutes, especially the women, because you need to understand one technical term for lesson. That term is *mitochondrial DNA*. You will be happy to know about this term if you don't already know.

---

1.   Book by Bryan Sykes, published by W. W. Norton, New York-London.

Mitochondrial DNA

There are two different versions of DNA in every human cell. One is the main DNA that is located in the nucleus of the cell and contains the genes from both the mother and the father. This is by far the larger of the two. The other is the mitochondrial DNA supplied only by the mother and is much smaller, but is less likely to change.

Mitochondria are a substance whose job is to help cells use oxygen to produce energy. The more vigorous the cell the more energy it requires and so the more mitochondria it needs. Cells from active tissue like muscle, nerve, and brain contain up to 1,000 mitochondria each.

Mitochondria are a unique version of DNA and it is inherited only from the mother. They are in the egg and there are lots of them; in comparison, the sperm has only a few mitochondria, just enough to provide energy for swimming up the uterus as they home in on the egg. Only the sperm head with its package of nuclear DNA enters the egg. The remaining mitochondria are jettisoned along with the tail. The plump fertilized egg now has a nuclear DNA from both parents and is ready to start dividing, but its only mitochondria are the ones that were in the egg all along and they all came from the mother. For that simple reason, the mitochondrial DNA is always maternally inherited.

The Iceman was found in 1991 in the Alps of Italy. It was necessary to determine his age. Bryan Sykes was called in because of his experience collecting DNA from ancient bones. They decided to use mitochondrial DNA for better results and they were successful. His age was estimated at 5,350 years and (German scientists were also asked to check at the same time, and the DNA results were identical.) Afterwards, Bryan ran checks on other DNA samples that were in their files and they found a match. She was an Irish friend of Bryan Sykes and she was living in Dorset a town in southern England. Her name is Marie Mosely and she had exactly the same DNA as the Iceman. This could only mean that there had to be an unbroken genetic link between Marie and the Iceman's mother, stretching back over five thousand years and faithfully recorded in the DNA. Since 1991 Dr. Sykes has written his book, "The Seven Daughters of Eve", and he has named the seven women based on where their bones were found and dated.

Let's look at the seven daughters of Eve. They are Ursula, 45,000 years old; Xenia, 25,000; Helena, 20,000; Velda, 17,000; Tara, 17,000; Katrine, 15,000; and Jasmine, 10,000. Bryan's team knows where the skeletons come from and they know the approximate age of each. He calls them clan mothers. For example the Iceman came from the clan Katrine. Dr. Sykes was besieged by request from people all over Europe and other countries as well who wanted to know who their ancestors were on their mother's side. For example: Brigette Bardot—clan Helena; Maria Callas—clan Ursula; Jennifer Lopez—clan Velda; Bryan Sykes—clan Tara; the Iceman, Marie Mosley, and me—clan Katrine.
(www.oxfordancestors.com)
(LUCA—Last Universal Common Ancestor is a term used to explain a link between species.)

Let's now go back to our lesson for today and remember that I ask you to remember that God refers to wisdom as "she" or "her", feminine words. When those words were written only God knew about mitochondria.

Epilogue

"Everything is only for a day, both that which remembers and that which is remembered. Observe constantly that all things take place by change, and accustom thyself to consider that the nature of the universe loves nothing so much as to change the things that are and to make new things like them."

Marcus Aurelius (A. D. 121-18), Roman emperor (161-180), and Stoic philosopher.

What will the future have in store? Will we kill each other with wars and terrorist, will we succumb to bacteria and viruses. Will global warming cause a massive extinction, followed by another Ice Age, or will be struck by a large meteor, comet, etc. The one thing that is constant is that there will be change.

I prefer to hope that Jesus will return with the promise of an improved new species that has a new DNA equipped with new genes for protection against bacteria and viruses and that will full-fill the promise of a more peaceful way to get along with our fellow travelers.

In the meantime, could we learn to live together? God bless us all, everyone.

Reference

Adams, Noah   "The Flyers—The Search of Wilbur & Orville Wright" Crown Pub.

Alper, Joe   "Rethinking Neanderthals" Smithsonian Magazine, June 2003

Aurelius, Marcus   "Meditations" Translated by George Long (Barnes & Noble Books)

Bascomb, Willard   "Deep Water, Ancient Ships" Mediterranean & Black Sea

Darwin, Charles   "The Origin of the Species" Bantam Books—New York

Dawkins, Richard   "The Selfish Gene" Oxford University Press—Oxford, NY

Eldredge, Niles   "The Triumph of Evolution" W, H, Freeman & Company New York

Reference Book   "World Book Encyclopedia" 1961

Gilgamesh   "The Epic of Gilgamesh" Penguin Classic (Babylon Mythology)

Gould, Stephen Jay   "Full House" Three Rivers Press New York
"The Book of Life" An Illustrated History of the Evolution of Life on Earth. (Introduction-Life and Times by Michael Benton.)

Hahn, Harley   "Internet Advisor" How to Make Online Life Work for You. QUE

Hawkins, Stephen   "A Brief History of Time"
"The Universe In a Nutshell" Bantam Books, New York

Howard, W. J.   "Life's Beginnings" Coast Publishing, Coos Bay, Or.

Ingle, Larry H.   "First Among Friends" George Fox & the Creation of Quakerism Oxford University Press, Inc. New York, NY

Lambert, David & The Diagram Group   "The Field Guide to Early Man" Fact on File Publication—New York, NY—Oxford, Eng.

Maddox, Brenda   "Rosalind Franklin—The Dark Lady of DNA" Harper Collins

Mayr, Ernst   "What Is Evolution" Published by Basic Books, New York

McGrath, Alister   "In the Beginning" The Story of the King James Bible.
The Story of the King James Bible and How It Changed a Nation, a Language, and a Culture. Published by Anchor Books New York

Miller, K.R.   "Finding Darwin's God" Harper Collins Publisher, New York

Olson, Steve   "Mapping Human History" Houghton Mifflin Company, NY

Ryan, William B.F.   "Noah's Flood" The New Scientific Discoveries—Touchstone, NY
Pitman, Walter

Ridley, Matt   "Genome" Harper Collins Publisher, New York

Schroeder, Gerald L.   "The Science of God" Broadway Books—New York

Stwertka, Albert   "A Guide to the ELEMENTS" Revised Edition, Oxford Press

Sykes, Bryan   "The Seven Daughters of Eve" W. W. Norton, New York-London

Ussher, James   Archbishop & Theologian (Estimate of Creation) (1581-1656)

Watson, James   "The Double Helix" Simon & Schuster New York-London

Zimmer, Carl   "Evolution" A PBS Television Series

0-595-32727-3